Nelson Maths

This book belongs to:

_____

Workbook Starter C

Karen Morrison
Lisa Greenstein

OXFORD
UNIVERSITY PRESS

Great Clarendon Street, Oxford, OX2 6DP, United Kingdom

Oxford University Press is a department of the University of Oxford.

It furthers the University's objective of excellence in research, scholarship, and education by publishing worldwide. Oxford is a registered trade mark of Oxford University Press in the UK and in certain other countries.

British Library Cataloguing in Publication Data

Data available

ISBN: 978-1-382-01040-5

10 9 8 7 6 5 4 3

Paper used in the production of this book is a natural, recyclable product made from wood grown in sustainable forests. The manufacturing process conforms to the environmental regulations of the country of origin.

Printed in India by Multivista Global Pvt. Ltd

Acknowledgements

The publisher and authors would like to thank the following for permission to use photographs and other copyright material:

Cover: Matthieu Nivesse.

Artwork by Aviel Basil, Andy Peters, Pantek Media, OKS Prepress, and Integra Software Services.

Every effort has been made to contact copyright holders of material reproduced in this book. Any omissions will be rectified in subsequent printings if notice is given to the publisher.

The manufacturer's authorised representative in the EU for product safety is Oxford University Press España S.A. of El Parque Empresarial San Fernando de Henares, Avenida de Castilla, 2 – 28830 Madrid (www.oup.es/en or product.safety@oup.com). OUP España S.A. also acts as importer into Spain of products made by the manufacturer.

Contents

Counting to 12

Draw and count

Draw 1 more in each set.

Write how many in each set.

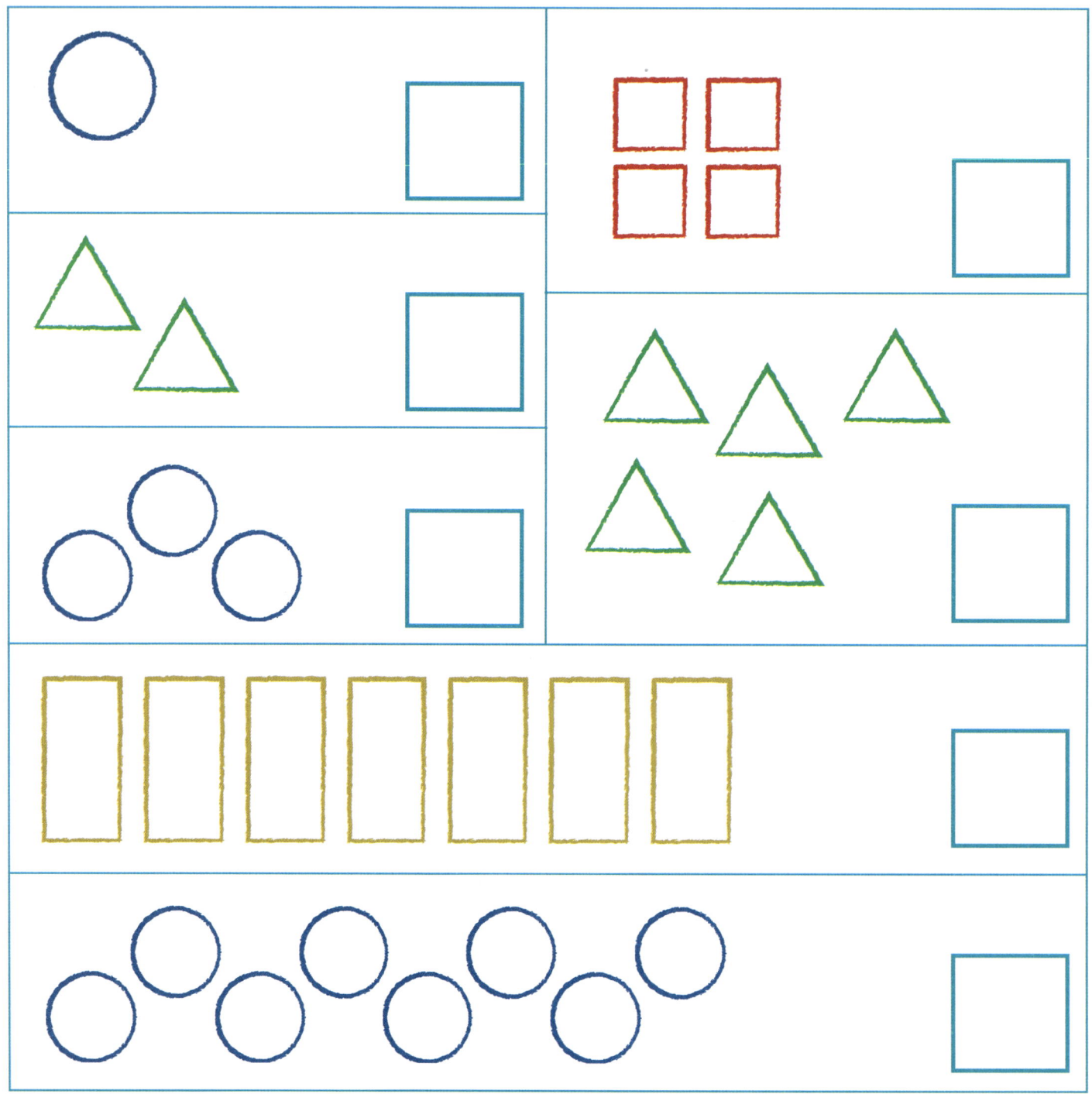

Sets

Draw the fruits in the correct sets.

banana   strawberry   lemon   apple   watermelon

Yellow fruits	Red fruits

For each fruit, circle the set that has more.

11 eleven

There are 11 leaves. Count to 11.

| 1 | 2 | 3 | 4 | 5 | 6 | 7 | 8 | 9 | 10 | 11 |

Draw 11 leaves on the plant.

Cross out ⌧ the sets that **do not** have 11 shapes.

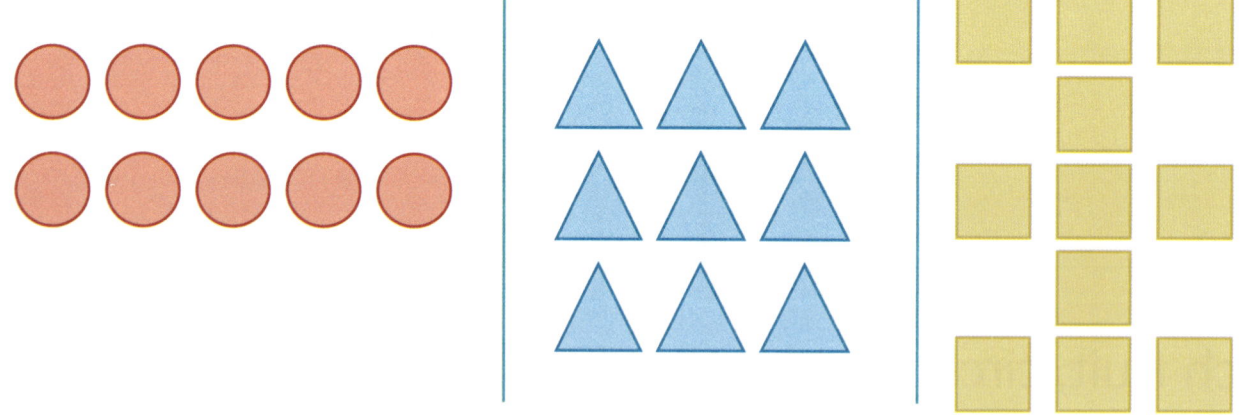

Write 11 and the word eleven.

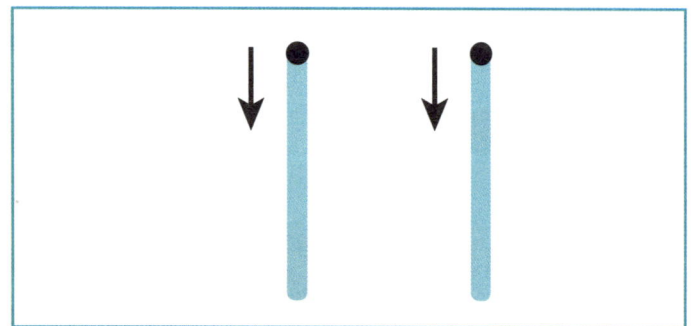

12 twelve

There are 12 fingerprint bugs. Count to 12.

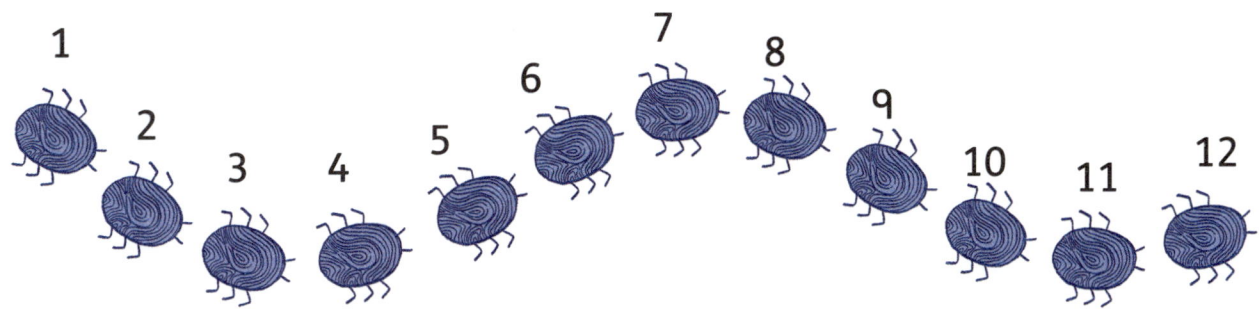

Write 12 and the word twelve.

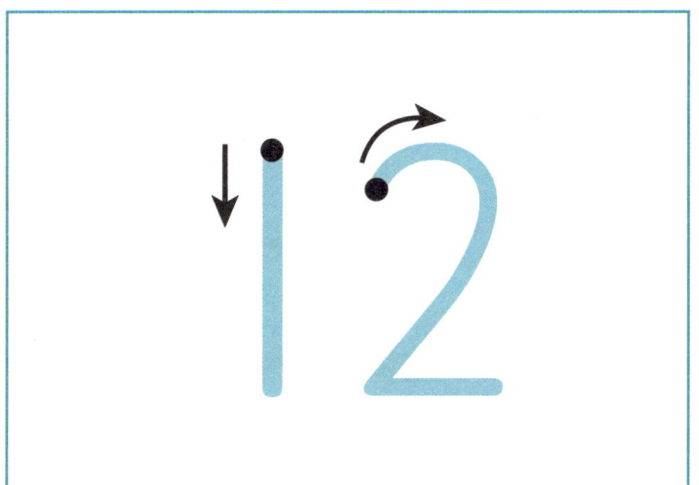

Use your own fingerprints to make 12 bugs.

Time

The short hand shows the **hour**.

3 o'clock

What time is it? Write the time.

☐ o'clock

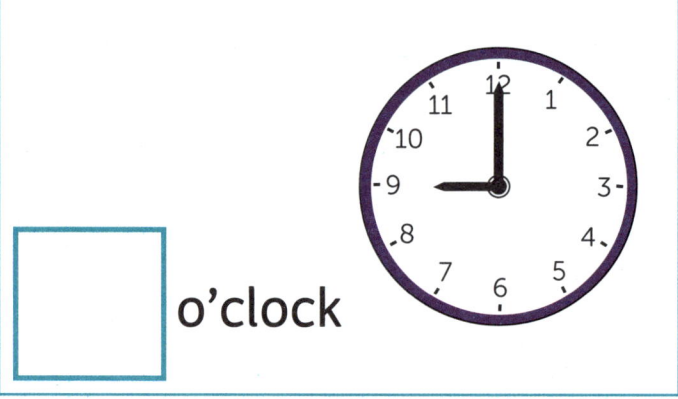

☐ o'clock

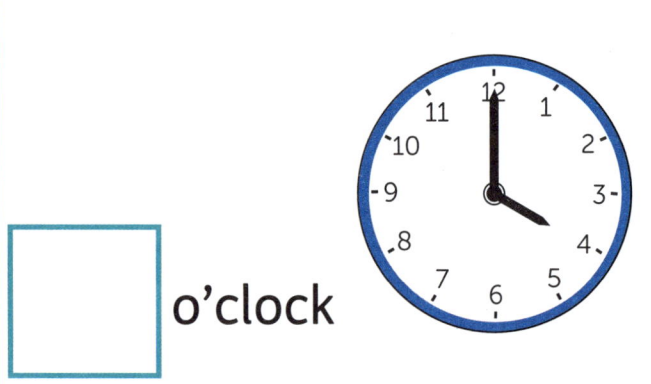

☐ o'clock

☐ o'clock

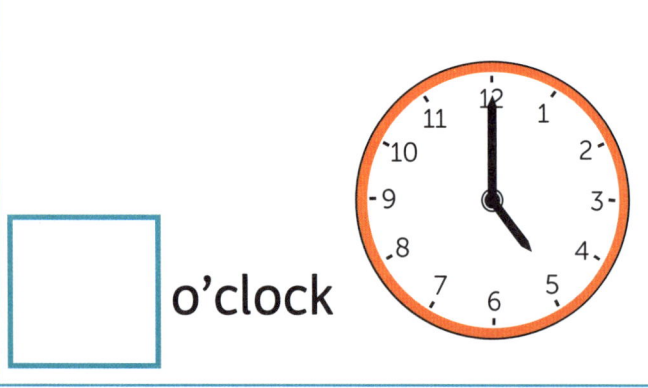

☐ o'clock

☐ o'clock

Show the time

At an o'clock time, the long hand points to 12.

3 o'clock

Draw the short hand to show the hour.

4 o'clock

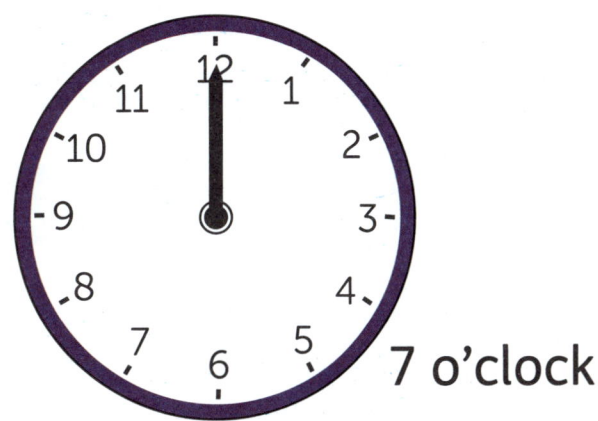

7 o'clock

1 o'clock

10 o'clock

2 o'clock

6 o'clock

Day and night

day   night

Does each picture show day or night?

Draw a sun  or moon .

Which takes longer?

Talk about which takes longer.

Tick ✓ the one that takes longer.

brushing your teeth ☐ or travelling to school? ☐

playing sports ☐ or tying your shoelace? ☐

eating breakfast ☐ or sleeping? ☐

What happens first?

Match the pairs of pictures.

Write 1 next to the thing that happens first.

Write 2 next to the thing that happens second.

One is done for you.

peel it

1

stick it

go to sleep

get dressed

wake up

go outside

cut it out

eat it

2

Tables

Inside or outside the box

Is it **inside** or **outside** the box?

Tick ☑ to show.

One is done for you.

	Inside	Outside
ball	✔	
book		
cup		
bear		
brush		
block		

Inside or outside the house

inside   outside

Do we use it **inside** or **outside** the house?

Tick ☑ to show.

One is done for you.

	Inside	Outside
football		✔
umbrella		
lamp		
rug		
bike		
bed		
car		

Sorting

Does it have corners or no corners? Is it plain or patterned?

Draw the shapes or write their letters.

One is done for you.

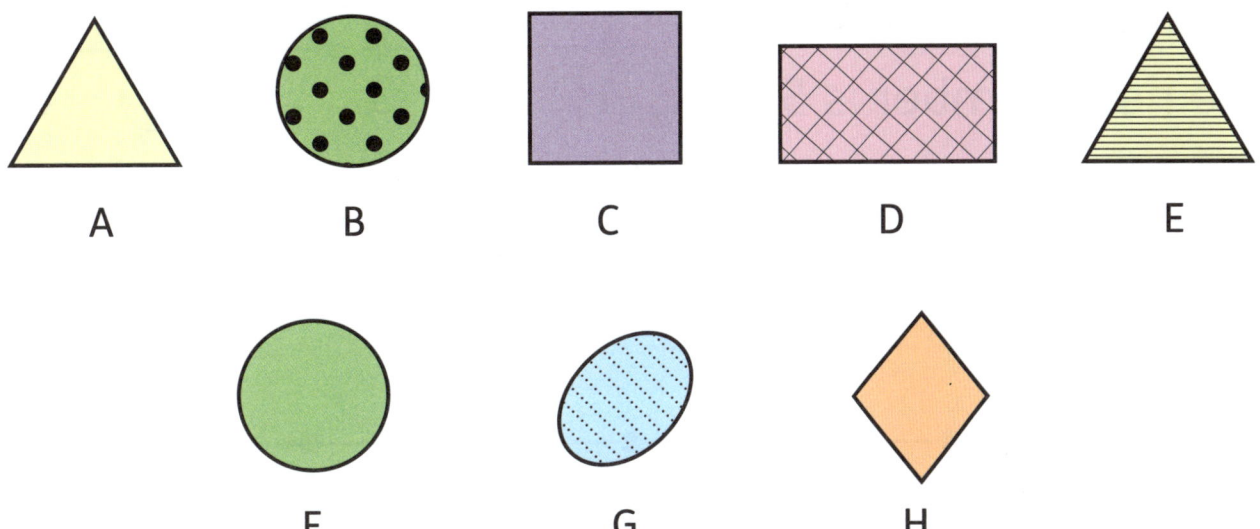

A   B   C   D   E

F   G   H

	Corners	No corners
Plain		
Patterned		B

Counting and doubling

Count forwards

The rabbit starts at 2.
It jumps 3 forwards.
The new number is 5.

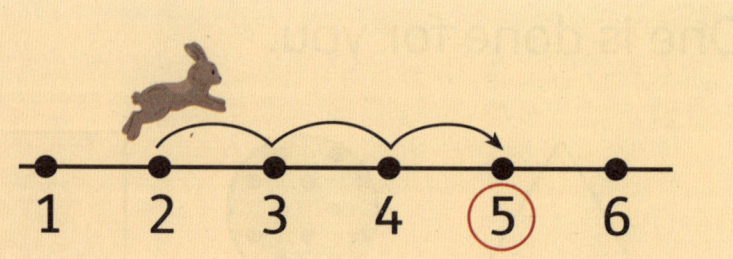

Count the jumps. Circle the new number.

Start at 0. Jump 3 forwards.

Start at 5. Jump 2 forwards.

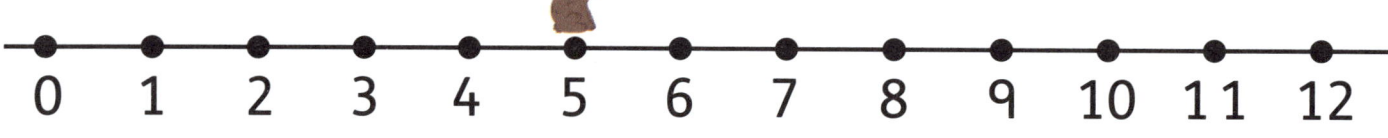

Start at 6. Jump 4 forwards.

Start at 4. Jump 6 forwards.

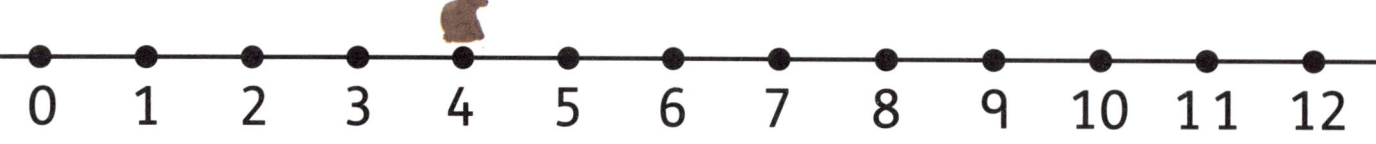

Start at 7. Jump 4 forwards.

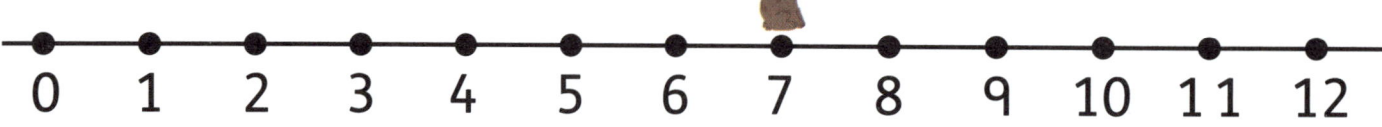

Count backwards

The rabbit is going back.

It starts at 6.

It jumps back 2.

The new number is 4.

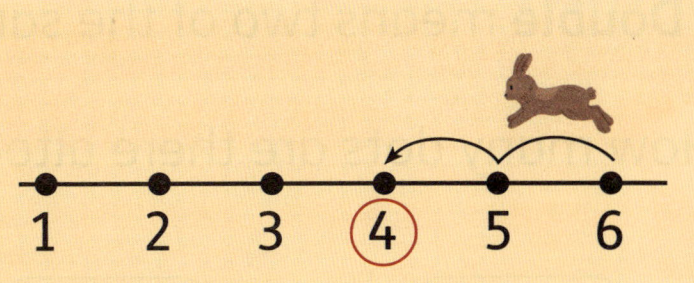

Count the jumps. Circle the new number.

Start at 2. Jump 1 back.

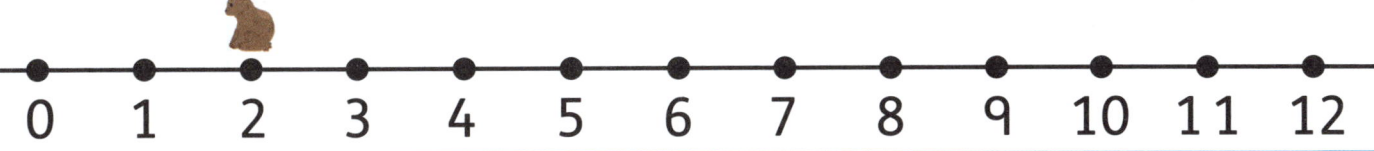

Start at 5. Jump 3 back.

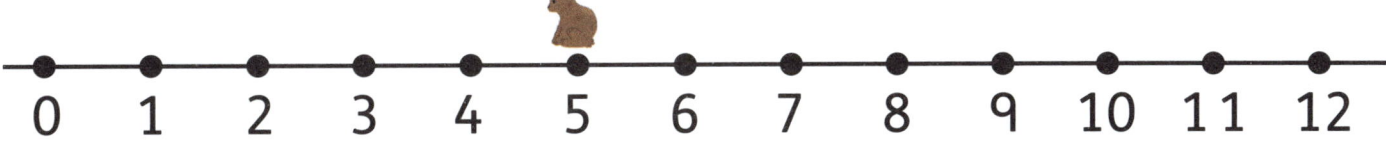

Start at 6. Jump 4 back.

Start at 10. Jump 6 back.

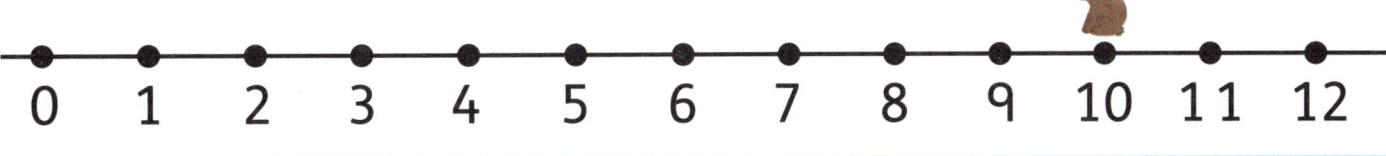

Start at 7. Jump 7 back.

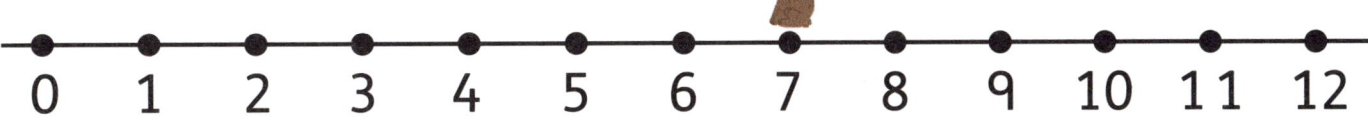

Double means two of the same number.

How many dots are there altogether?

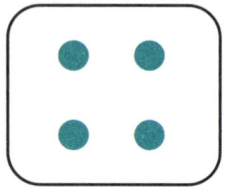 +    double 4 =

 +    double 1 =

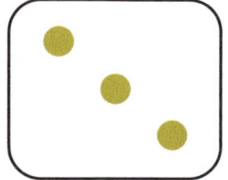 + 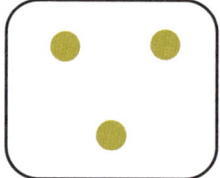   double 3 =

 + 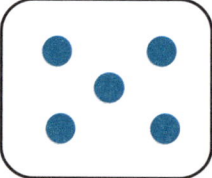   double 5 =

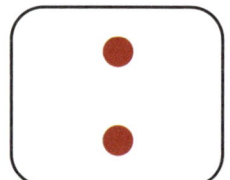 + 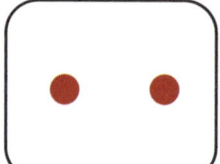   double 2 =

 +    double 6 =

Money

Money in coins

Find four different coins.

Put the coins under a piece of drawing paper.

Rub the coins with a crayon and cut out the shapes.

Stick a different coin picture on each jar.

Money in notes

Draw some notes from your country.

Add money

In Britain, people pay with money called pounds (£).

£2 + £1 = £3

£2   £1

Write how many pounds.

£2 + £2 = ☐

£10 + £1 = ☐

£2 + £3 = ☐

£5 + £1 = ☐

£2 + £10 = ☐

£5 + £5 = ☐

UNIT 6

Sharing

Share between 3

Three friends share 12 cookies. They all get the same number of cookies.

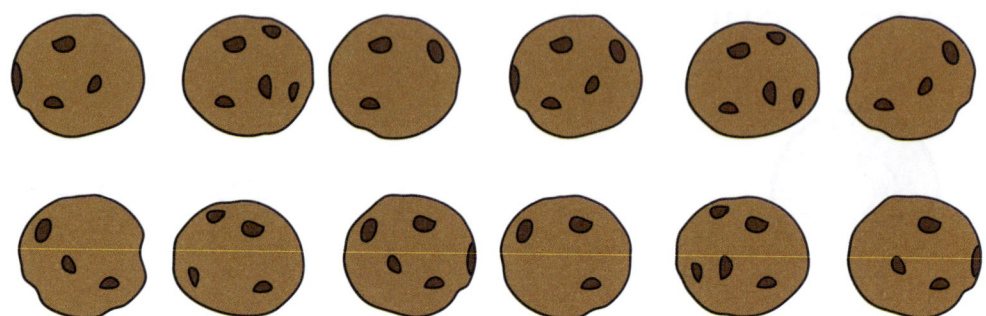

Draw how many each child gets.

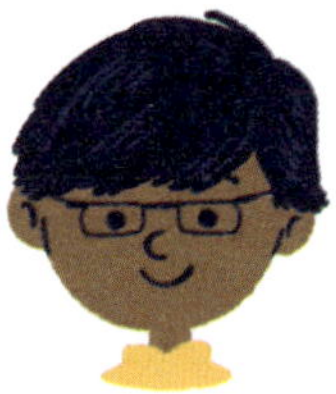	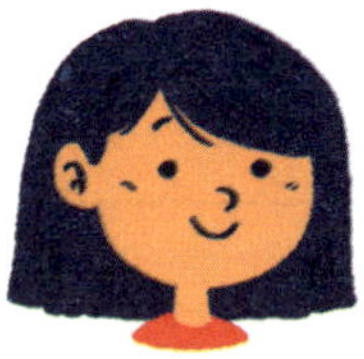	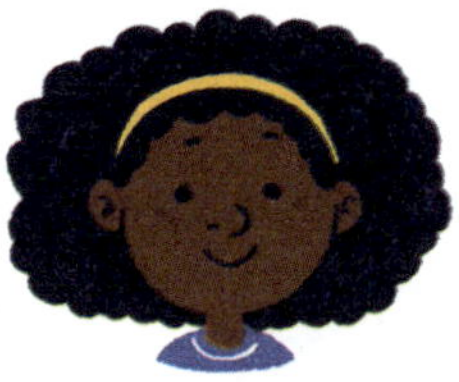

Share between 4

Four friends share these stickers. Each friend gets the same number.

Count the stickers. Draw how many each child gets.

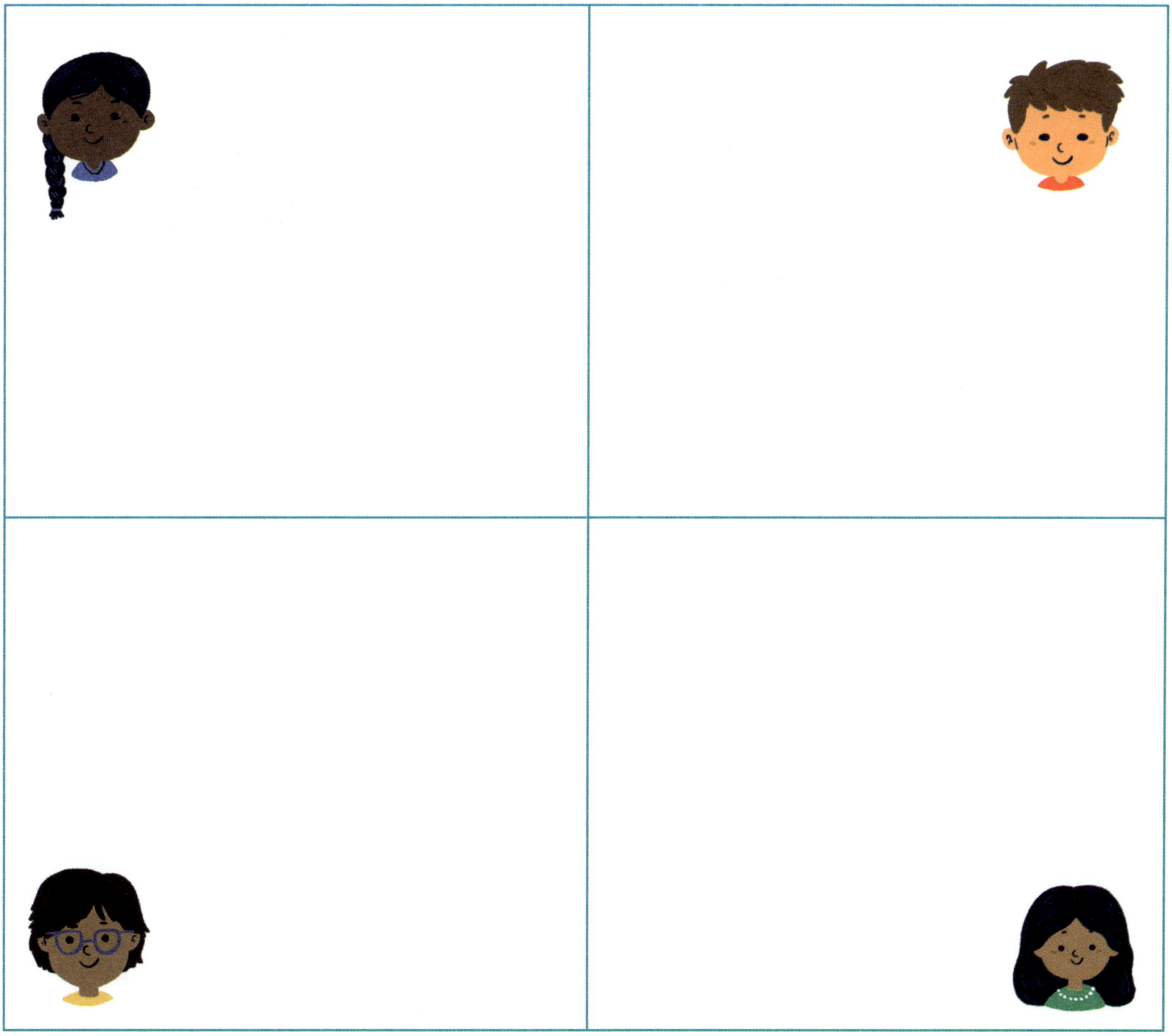

More sharing

Share fairly.

Write how many each child gets.

sweets

flowers

balloons

sweets

flowers

balloons

sweets

flowers

balloons

UNIT 7

Patterns

Make patterns

Continue these patterns.

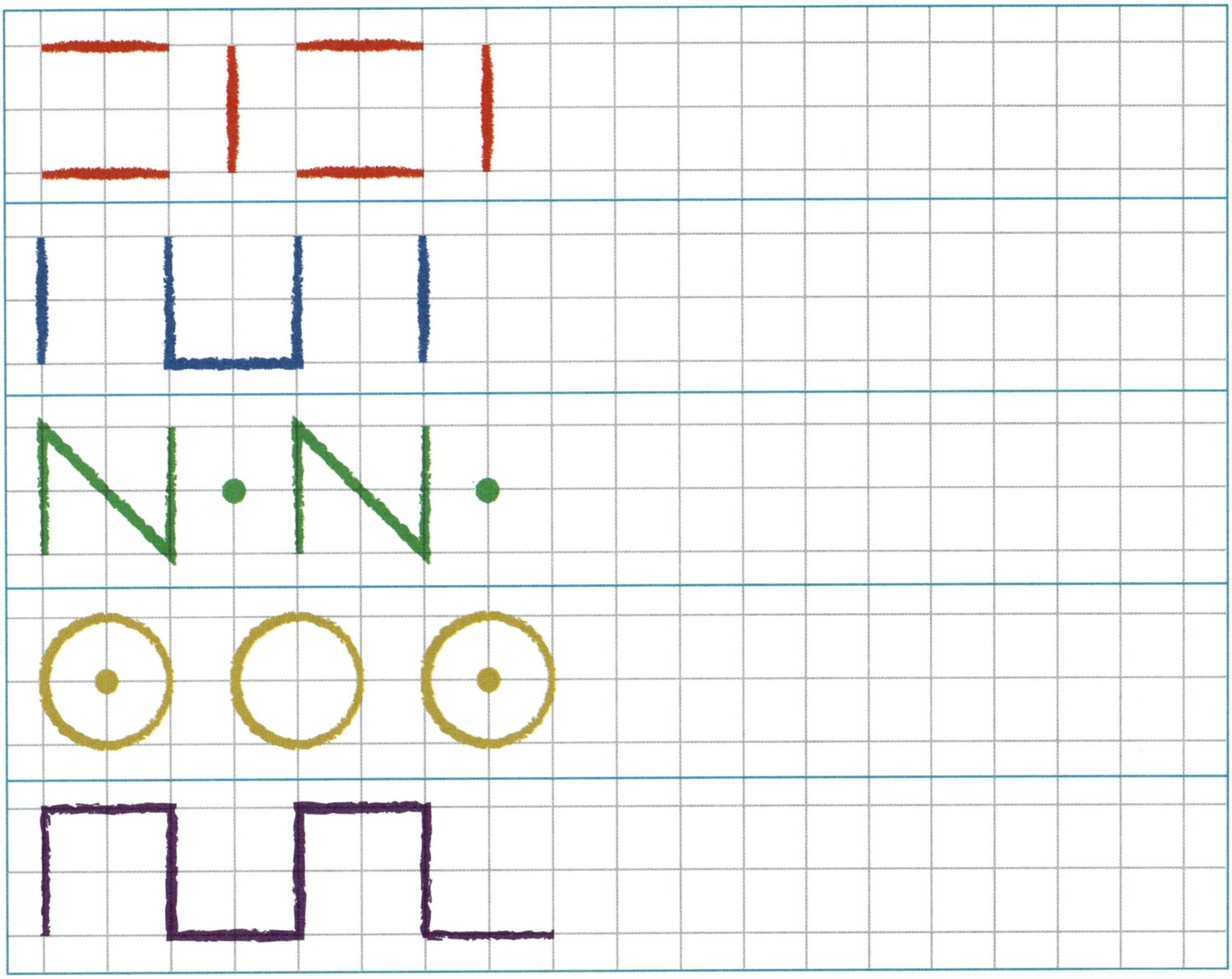

Now make your own patterns.

13 thirteen

10 and 3 makes 13.

Draw dots on 13 triangles.

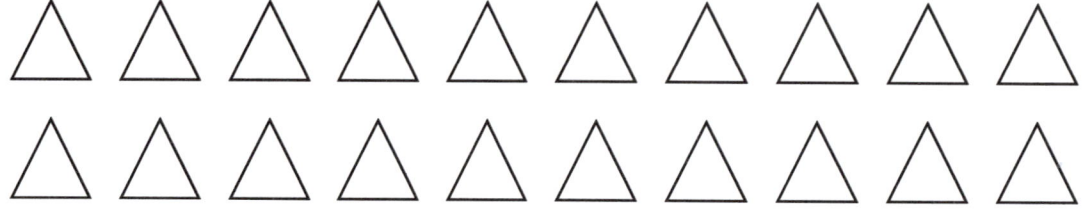

Write 13 and the word thirteen.

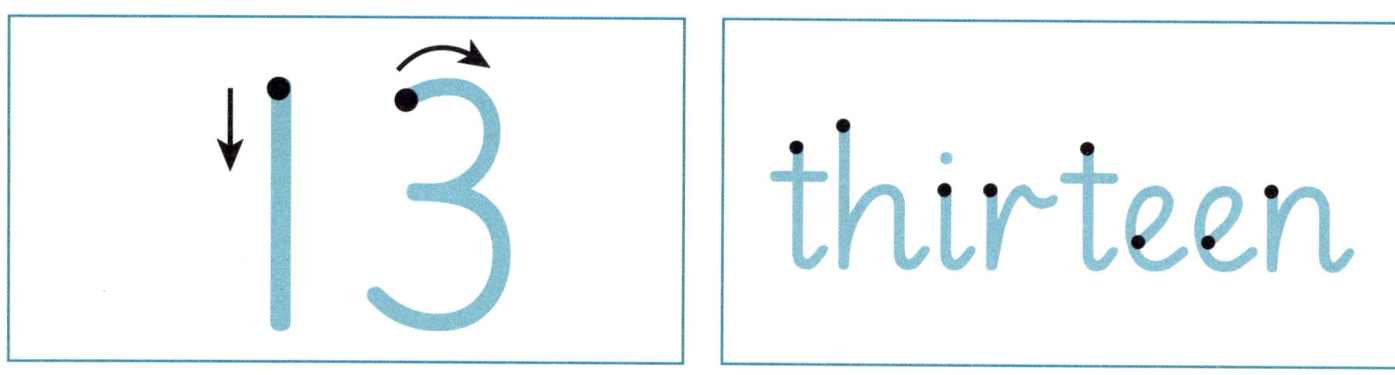

Cross out ☒ the sets that **do not** have 13 shapes.

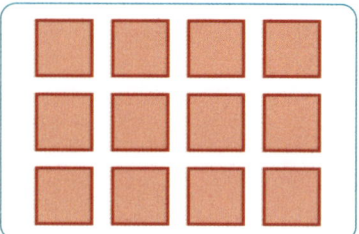

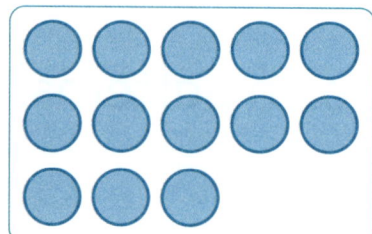

14 fourteen

10 and 4 makes 14.

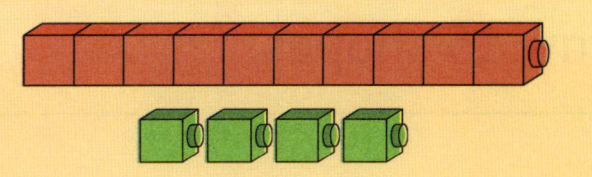

Draw dots on 14 stars.

Write 14 and the word fourteen.

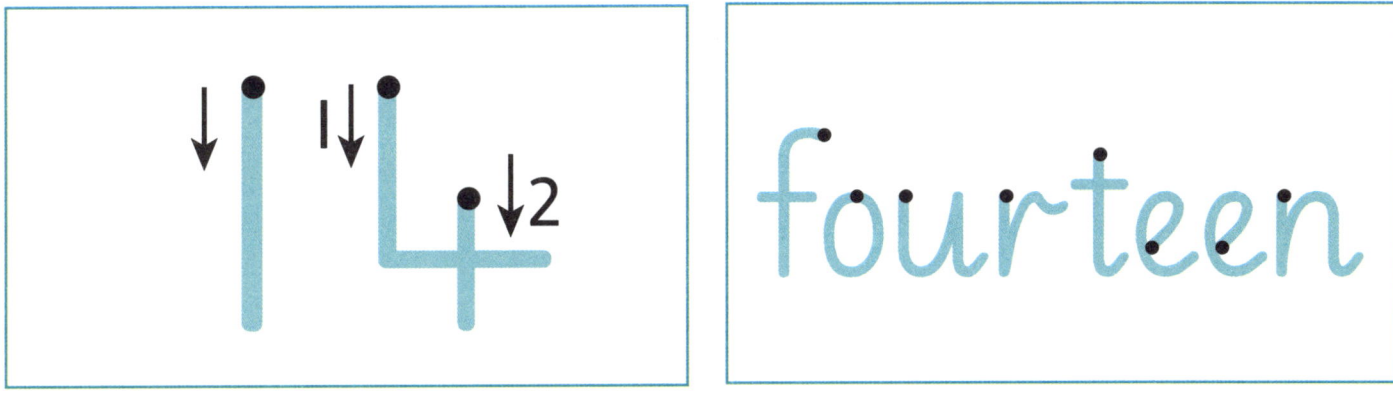

Draw lines to share 14 into two **equal** groups.

How many dots in each group?

Draw 15 shapes.

Complete: 10 and ☐ makes 15.

Write 15 and the word fifteen.

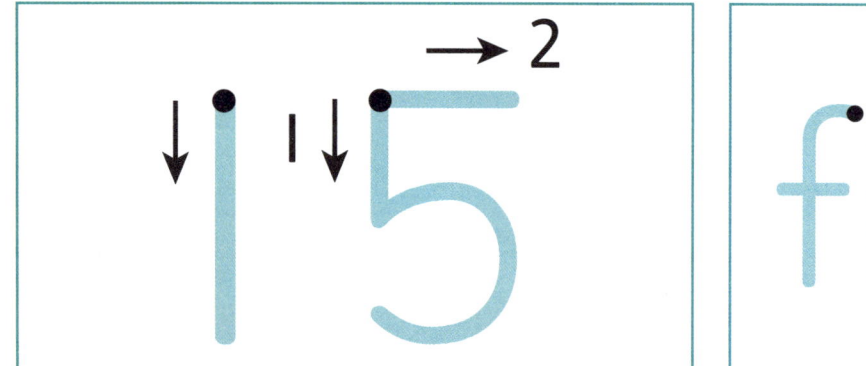 

Draw lines to share 15 into three **equal** groups.

How many dots in each group?

Working with numbers

Missing numbers

Write the number that is 1 more.

| 1 | |

| 5 | |

| 9 | |

| 12 | |

Write the number that is 1 less.

| | 3 |

| | 8 |

| | 11 |

| | 15 |

Write the number that comes between.

| 8 | | 10 |

| 11 | | 13 |

| 10 | | 12 |

| 13 | | 15 |

How many seeds?

Write how many seeds altogether.

Colour the fruits.

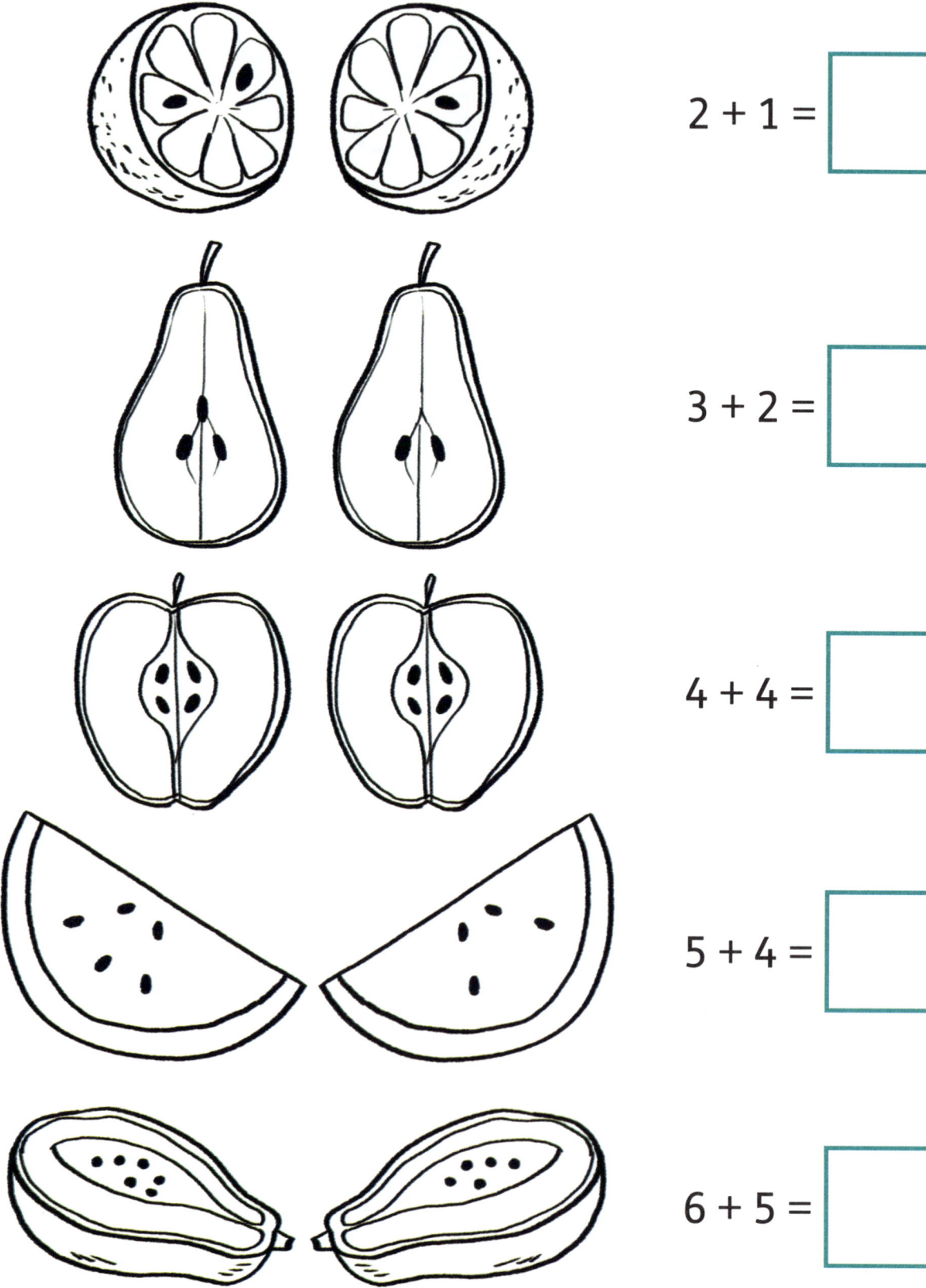

$2 + 1 =$ ☐

$3 + 2 =$ ☐

$4 + 4 =$ ☐

$5 + 4 =$ ☐

$6 + 5 =$ ☐

More than 10

Add to make numbers bigger than 10.

One is done for you.

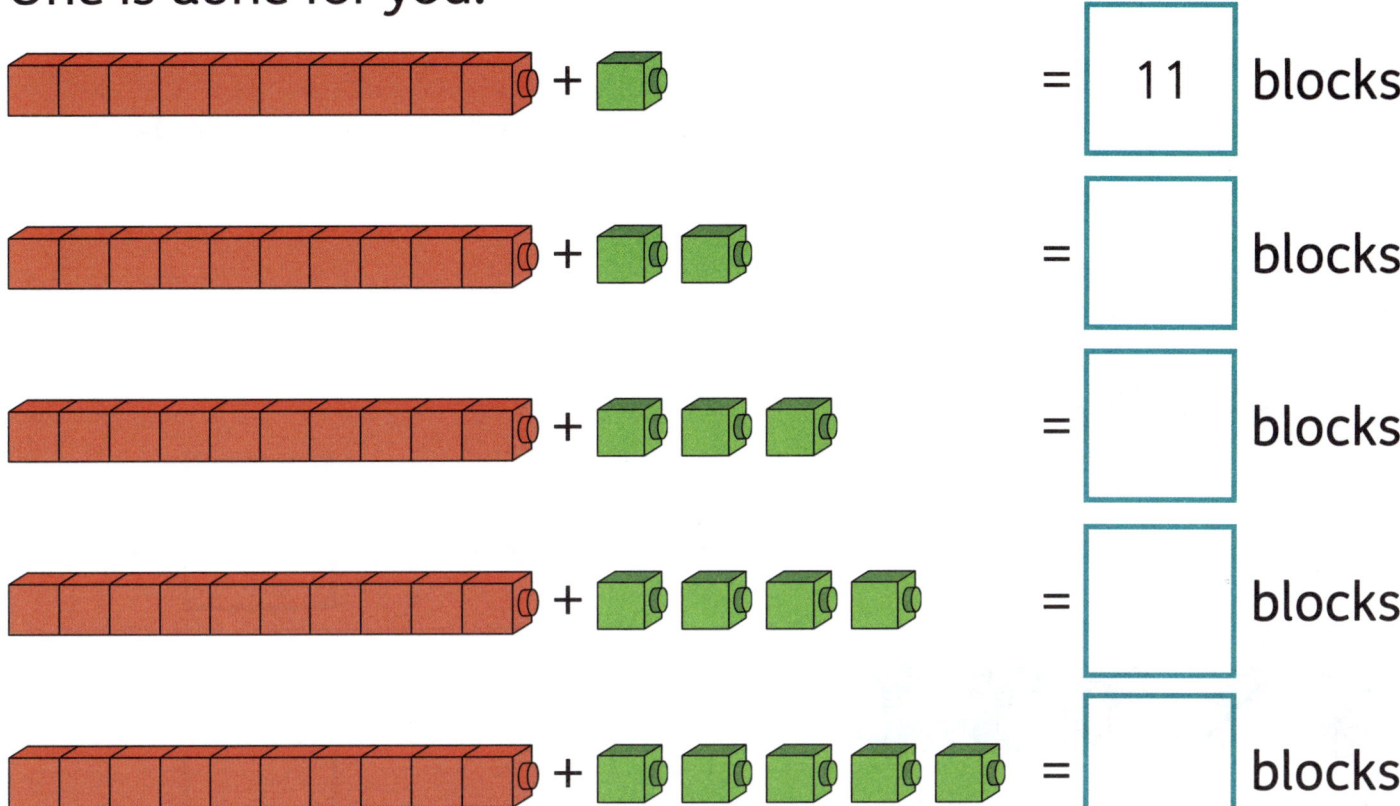

Write the numbers.

Draw lines to join each number to the correct picture.

twelve	thirteen
fourteen	fifteen

Take away

We can take away to make smaller numbers.

5 blocks. Take away 1.   4 left.

Write how many are left.

3 blocks. Take away 2. ☐ left.

5 blocks. Take away 2. ☐ left.

6 blocks. Take away 3. ☐ left.

7 blocks. Take away 2. ☐ left.

8 blocks. Take away 3. ☐ left.

How many are left?

Cross out  the objects.

Write how many are left.

One is done for you.

 Take away 2. `4`

 Take away 1.

 Take away 3.

 Take away 2.

 Take away 4.

Temperature

Hot or cold

Choose a colour for **hot** things and a different colour for **cold** things.

Colour them.

Counting to 17

16 sixteen

10 and 6 makes 16.

Write 16 and the word sixteen.

Draw 16 seeds on the watermelon.

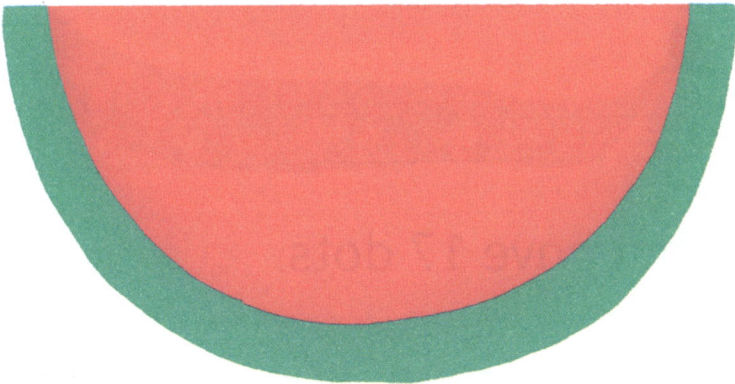

Circle the sets that have 16 dots.

10 and 7 makes 17.

Write 17 and the word seventeen.

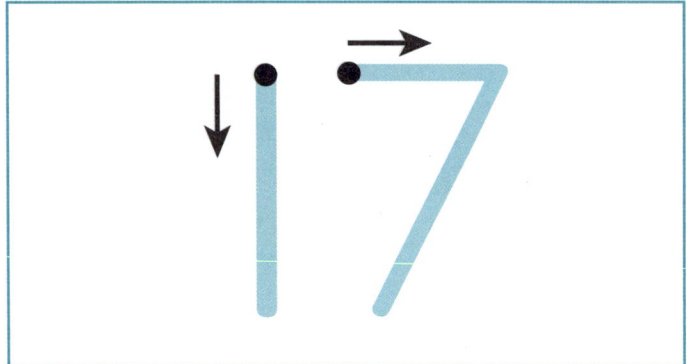

Draw more seeds on the fruit to make 17 seeds.

Circle the sets that have 17 dots.

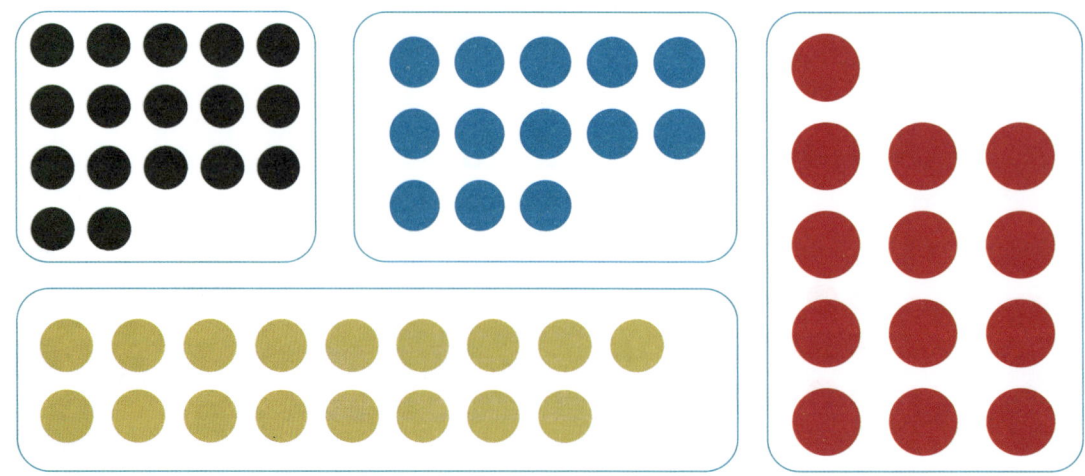

Up or down

Circle the one that is **up** in each picture.

Circle the one that is **down** in each picture.

Positions

Draw:

- a banana **above** the bread

- an apple **below** the jam

- some carrots **next to** the bread

- a cookie **next to** the eggs.

The bird must go up.

Colour the new position.

One is done for you.

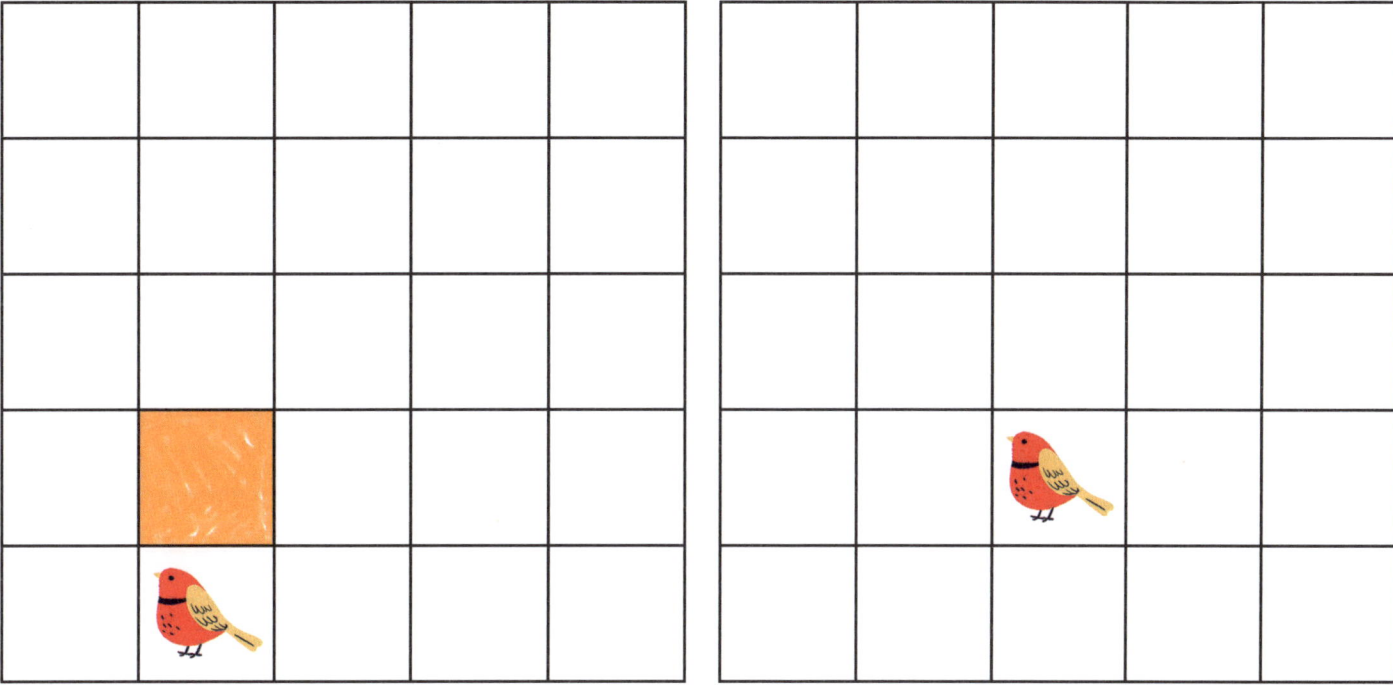

Go up 1 square.   Go up 2 squares.

Go up 3 squares.   Go up 4 squares.

Down

The raindrop must go down.

Colour the new position.

One is done for you.

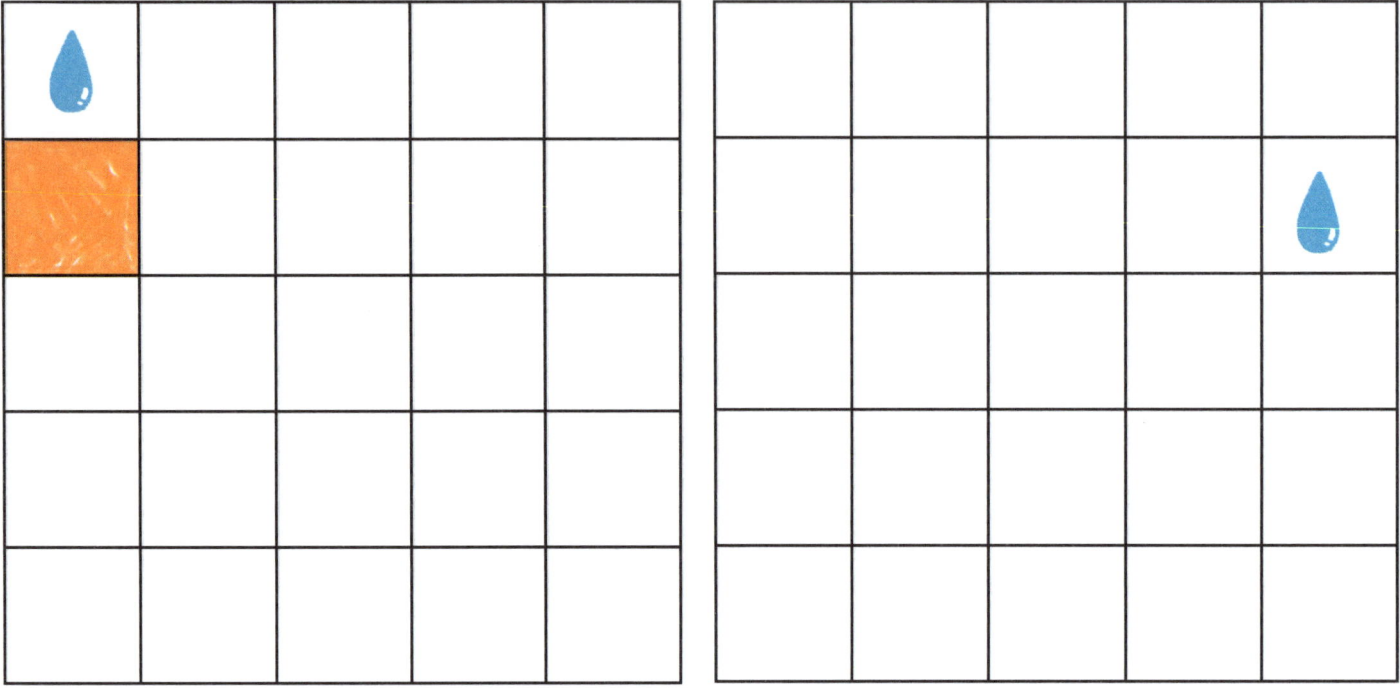

Go down 1 square.   Go down 2 squares.

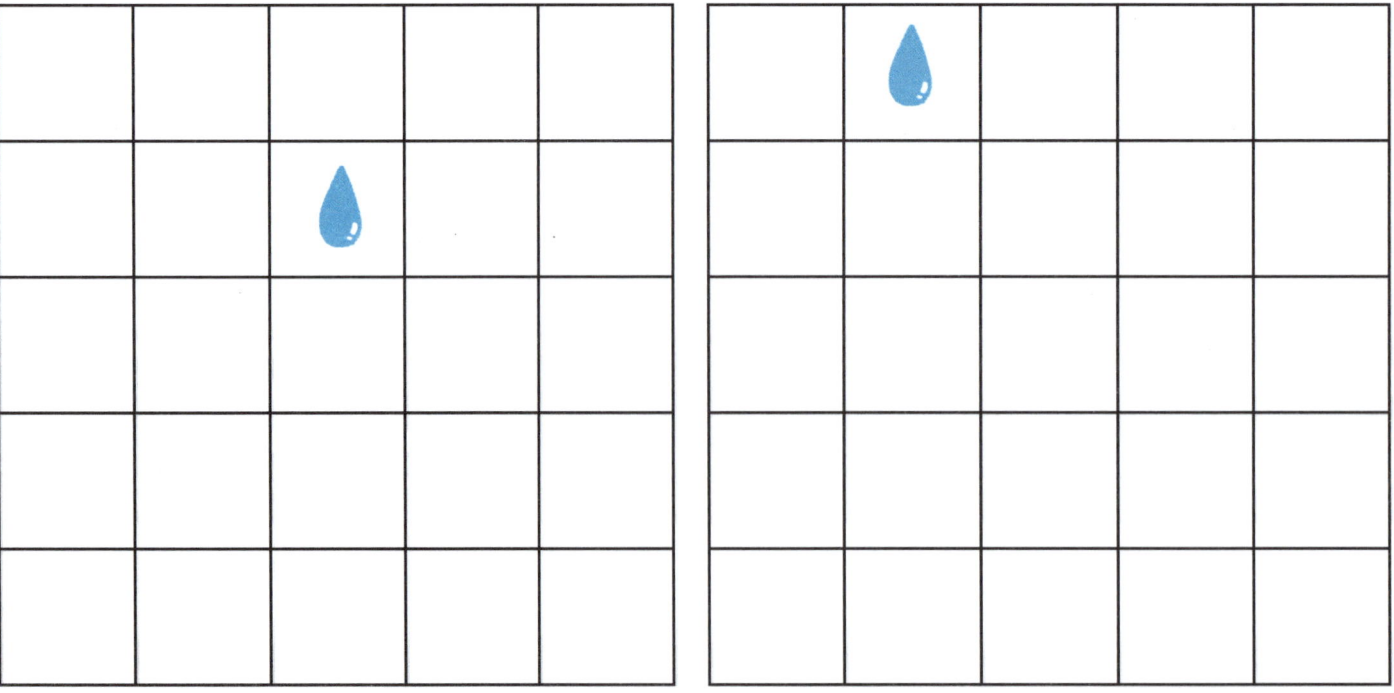

Go down 3 squares.   Go down 4 squares.

Simple graphs

Big and small

Draw 3 big leaves and 4 small leaves.

Colour the boxes to show how many of each size.

Small					
Big					

Different colours

Colour some fish green and some fish yellow.

Tick ☑ boxes to show how many of each colour.

Green							
Yellow							

Counting to 20

18 eighteen

10 and 8 makes 18.

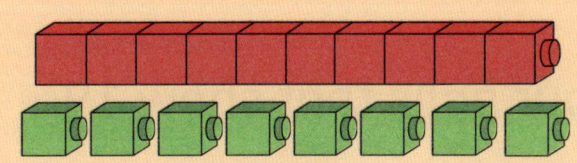

Tick ✓ the sets of 18.

 ☐

 ☐

 ☐

 ☐

Write 18 and the word eighteen.

Draw more dots to make a set of 18.

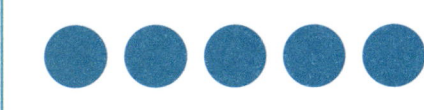

10 and 9 makes 19.

Colour balls that show 19.

Write 19 and the word nineteen.

Draw more dots to make a set of 19.

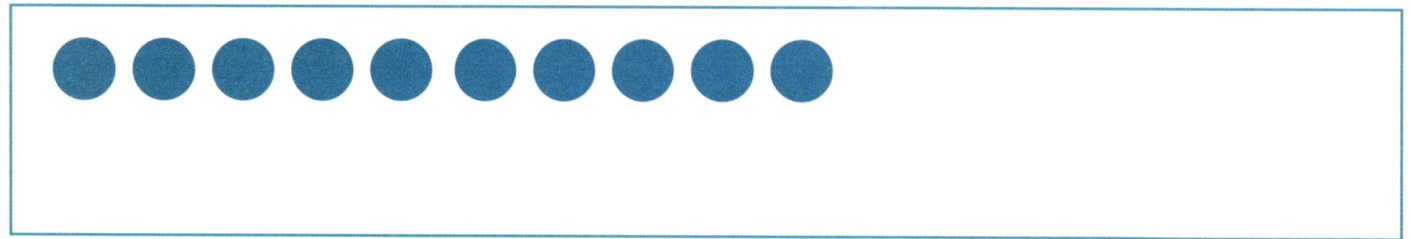

20 twenty

Tick ☑ the sets of 20.

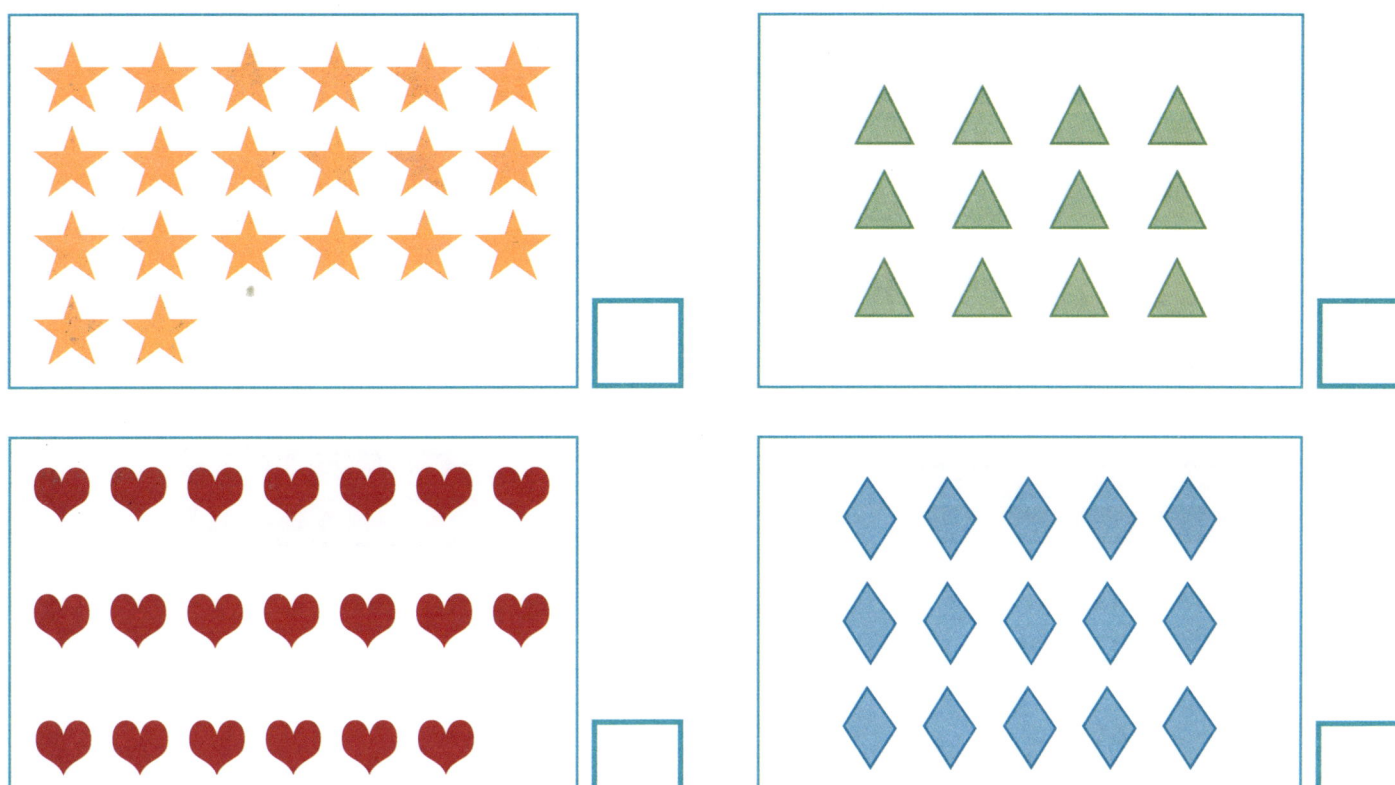

Write 20 and the word twenty.

Draw more dots to make a set of 20.

Now count as high as you can after 20.

Count in tens

Two tens are 20. We say twenty.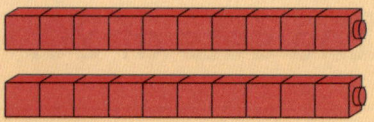

Count in tens.

Write the numbers and words.

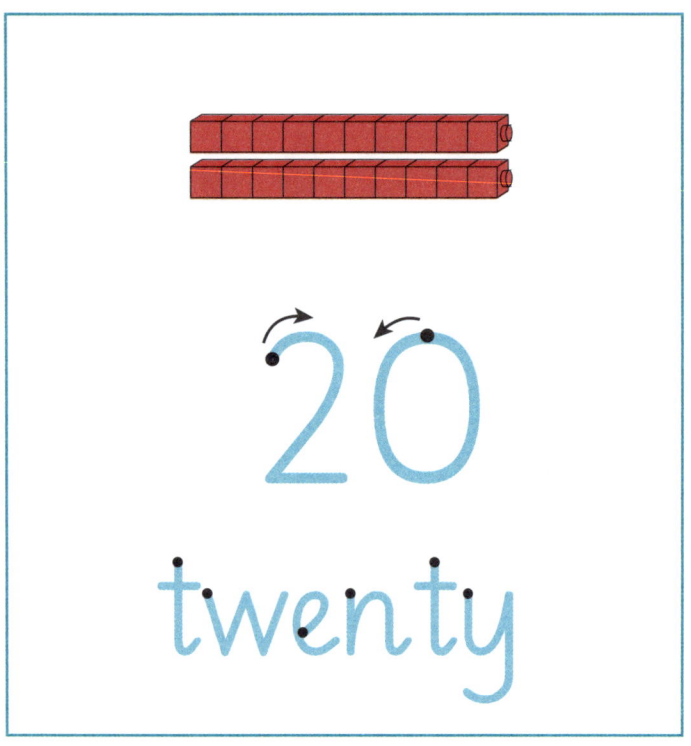

20 twenty

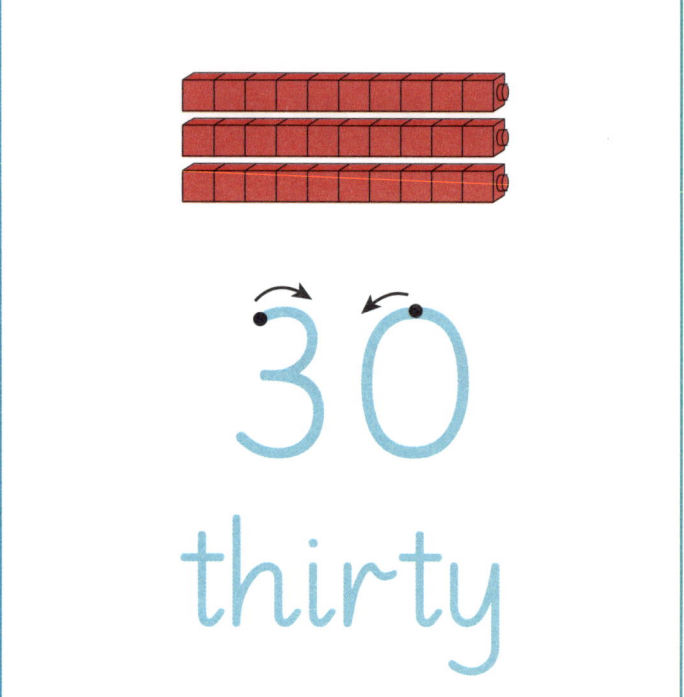

30 thirty

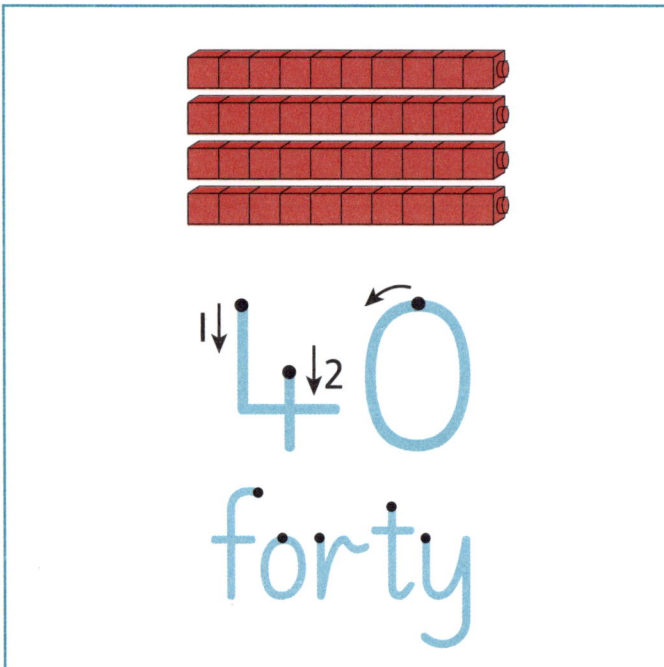

40 forty

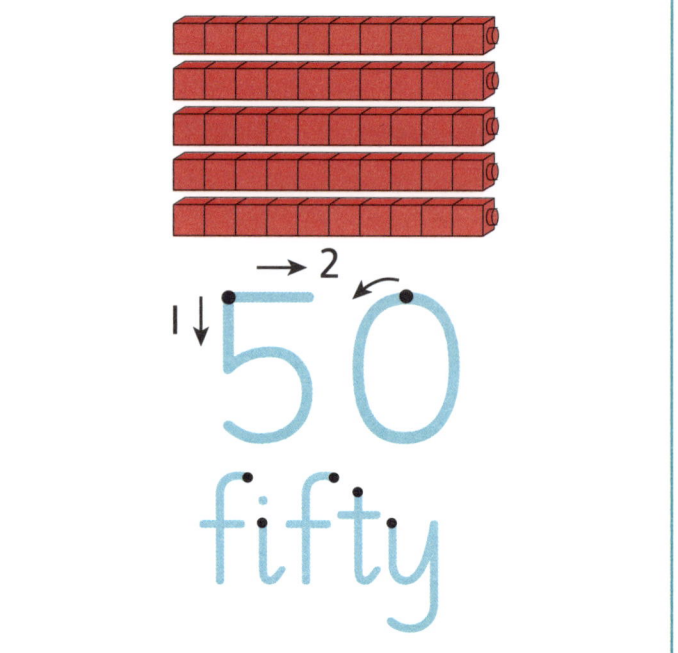

50 fifty

Bigger numbers

Circle the correct number in each row.

Twenty-one	20	21	22	23	24	25
Twenty-two	20	21	22	23	24	25
Twenty-three	20	21	22	23	24	25
Twenty-four	20	21	22	23	24	25

Write how many blocks.

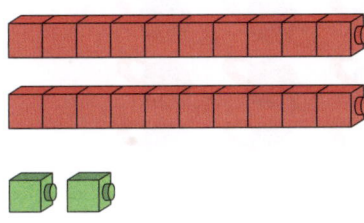  blocks.

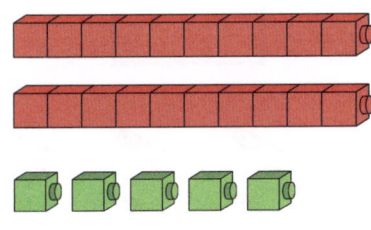  blocks.

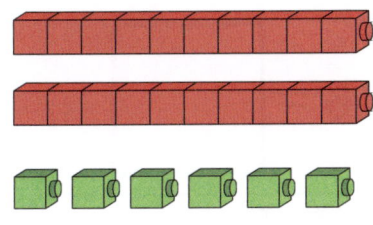  blocks.

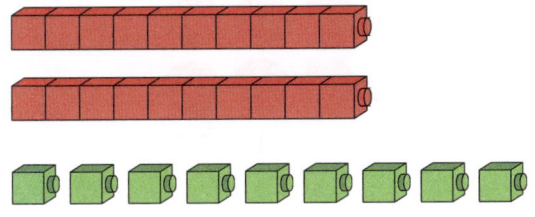  blocks.

Half of a group

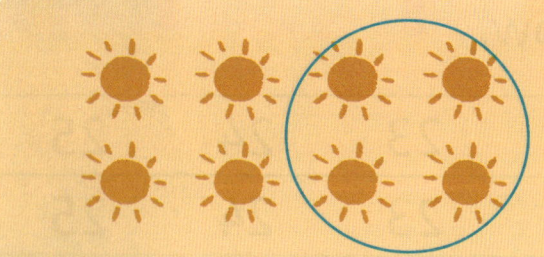 half of 8 = 4

Half

Draw lines to show half of each group. Write the number.

half of 10 = ☐

half of 16 = ☐

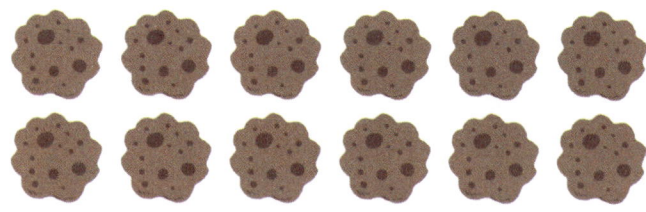

half of 12 = ☐

half of 18 = ☐

half of 14 = ☐

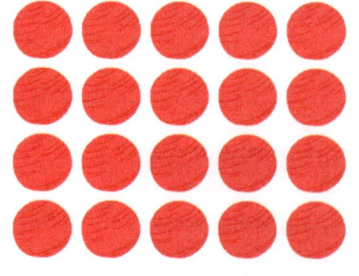

half of 20 = ☐